Ping-Ping the PANDA

Come on a great adventure with me and learn about my family

TRUE TO LIFE BOOKS

Educating children about endangered animals.

Jan Latta author and wildlife photographer.

Hello, my name is Ping-Ping the panda.
My home is a bamboo forest
on the mountains of China.

Giant pandas have lived in China for about three million years. Now we are one of the most **endangered** animals on earth.

This is my favourite tree stump.
I like to try many different positions
until I am comfortable. Then I fall asleep.

We make many different sounds to **communicate**. We cry like a baby, bark like a dog, or growl, squeal, or honk.

We love playing together and having a good wrestle. We like to see who can get to the top of the tree first.

When I was born I weighed only 90 grams.
Sometimes my mother gives birth to twins.
She will look after them for about two years.

My father weighs 120 kilograms. He can eat up to 16 kilograms of **bamboo** a day. Then he needs to rest so the bamboo can digest.

Bamboo is my favourite food.
There are 30 different types of bamboo.
I eat the leaves, stems and the tender roots.

I like to lie down or sit up when I'm eating so I can get a firm grip of the bamboo stems.

Sometimes I **eat** flowers, fruit, mushrooms and insects.

I walk or run on the ground with my padded **paws**.
I love climbing trees with my friends. It is great fun.

I like to **climb** up a tree and have a good scratch.
My thick fur helps to keep me snug and warm in winter.
Sometimes I curl up in a ball to keep warm.

See my little tail and five fingers on my **paws**. My front paws have a false thumb to hold bamboo stems.

I have a powerful jaw and strong teeth to crush and grind bamboo stems.
My eyes have slit pupils, just like a cat.

Meet my closest relative, the **red** panda.
They are also endangered. They live on the mountain slopes and love bamboo, just like me.

GIANT PANDA FACTS

The mountain areas where pandas can still be found.

SCIENTIFIC NAME

Ailuropoda melanoleuca (means 'black and white cat-footed'). Also known as large cat bear or bamboo bear.

HABITAT

Pandas live in the mountain regions of Gansu, Sichuan, and Shaanxi provinces in China.

WEIGHT

Males from 80 to 125 kilograms. Females up to 100 kilograms.

LENGTH

1.5 to 1.8 metres

BIRTH

Pandas weigh 84 to 140 grams at birth. (1/900th of their mother's weight). Pandas are pink and hairless at birth. The black patches appear after 10 days. They are born blind but will see after six to eight weeks. They will stay with their mother for about two years.

DIET

Pandas eat 30 different types of bamboo. The stalks, leaves and roots make up 95% of the panda's diet. They will also eat plants, fruit, flowers, mushrooms and insects. They spend 10 to 16 hours a day feeding. When feeding, they sit or lie down to get a better grip on the bamboo stems. Cubs need to eat five to seven times a day. Pandas usually drink water once a day.

PREDATORS

Humans are the only danger to pandas. They hunt them for their pelts.

LIFE SPAN

Pandas can live up to 20 years in the wild and 30 years in captivity.

NUMBERS REMAINING

There are approximately 2,000 pandas living in the wild in China. There are fewer than 150 living in zoos and breeding centres around the world.

DID YOU KNOW?

- Pandas were first sighted by Westerners in 1850.
- Wolong has the largest number of captive-bred pandas in the world.
- Pandas have a very good sense of smell.
- Pandas love having their backs scratched, just like a pet puppy.
- Panda babies are called cubs and cry just like a human baby.
- Pandas are good swimmers.
- Adult pandas live a solitary life.
- Pandas will roam over a large area covering a kilometre a day.
- Pandas do not hibernate like other bears.
- The giant panda is a symbol of peace and harmony in China.

CREATING PING-PING THE PANDA BOOK

“When I read about 16 baby pandas who were born at Wolong in China, I knew I had to find a way to photograph and write about them.

I went to Chengdu, hired a Chinese driver and interpreter, and for four hours we travelled up the mountains to Wolong. It was a terrifying trip, but worth it to see the pandas. I asked permission to play with them and it was allowed, provided I covered my shoes and clothes with plastic. The pandas pulled everything off and tried to climb my legs. They were very funny with their antics. It was a privilege to be so close to these beautiful endangered animals.”

Jan Latta, author and wildlife photographer.

QUESTIONS

1. Where do pandas live?
2. What colour is a panda when it's born?
3. What do pandas eat?
4. Are pandas related to bears?
5. How many sounds can a panda make?
6. Do pandas live alone?
7. When were pandas first seen on Earth?
8. Can pandas give birth to twins?
9. Can pandas swim?
10. Is a red panda related to the giant black and white panda?

THOMAS HAMLYN-HARRIS

FUN ANIMAL ACTIVITIES

PANDA PLAYTIME

Turn a plain paper bag upside-down to wear as a panda puppet over your hand. Decorate the bag to look like a panda. Draw on a face. Cut out ears and paws and glue them on.

Or make a panda face using a paper plate. Paint the face. Cut out ears, nose and eyes and glue them on. Pick bamboo branches. Or pick thin branches and make leaves with green paper, so you can pretend to be a panda in a bamboo forest.

ELEPHANT MASK

From one side of a cereal box cut out the shape of an elephant's head. Paint it grey or brown. Cut out tusks and ears and glue them on. Make a hole at either side and loop with string or elastic to wear.

ANIMAL ANCESTORS

Research and list the names of animals that were alive during the Ice Age. A woolly ancestor of the elephant roamed the earth. What was its name? Draw its picture. Why did it become extinct?

CHIMP CHOPSTICKS

Head outdoors, crouch down and use a wooden chopstick as a chimp's tool to forage for food in the ground. Record what you see at chimp level.

TRACK A TIGER

What are pug marks? Make a set of pug marks with a potato print. Cut a potato in half. Cut out the shape of the pug marks. Dip in paint and make a print of the tiger tracks on paper.

HAPPY LION FACE

On a piece of paper draw a lion's face and paint it yellow. Cut out ears and glue on. Mark eyes, nose and mouth in black. With glue, stick pasta spirals around the head for a mane.

KOALA PORTRAIT

Make a koala portrait using a paper plate. Paint the plate grey. Cut out ears, nose and eyes and glue them on. Pick gum branches and hang them in your classroom so your koalas can live in the trees.

GUMLEAF PRINTS

Collect gum leaves and make a paper rubbing of the leaves. Place a flattened leaf under a piece of paper and gently rub a crayon over the top, making sure you don't move the paper.

ORANGUTAN ORANGES

Make an orangutan face on an orange. Place each orange on a flat surface. If it's wobbly ask the teacher to cut a base. Stick on sultanas for eyes and nose and draw on a smile. Use strips of orange paper for its long hair.

GIRAFFE DOT PAINTING

Draw the outline of a giraffe on paper. Use a paintbrush to make an Aboriginal-style dot painting with brown, orange and yellow paints for its coat. Draw on a face with a marker, then hooves and a mane.

MEERKAT FINGER PUPPETS

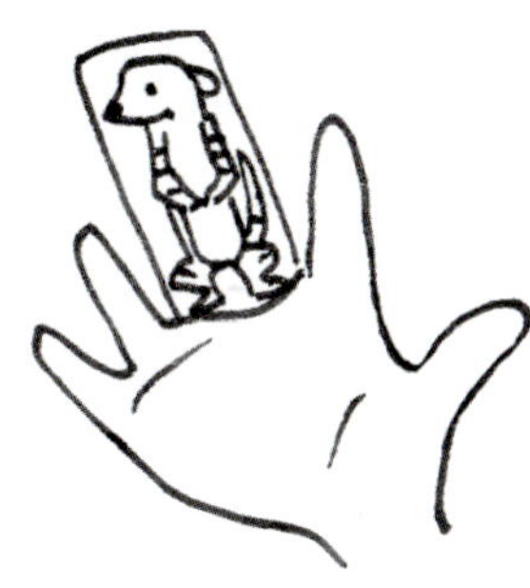

Place two pieces of paper flat in front of you and put your finger on top. Draw around the outline of your finger and cut out the shape allowing at least a centimetre extra space. Staple the top and sides together keeping the bottom edges open. Draw the face, arms, legs and add a tail.

RHINO HORN RATTLE

Make a rhino horn with black paper. Take an A4 size piece of paper length-ways and fold the two top corners to the middle and keep folding until you get a cone/horn shape, pointed at one end. Tape the horn shape and fill with a few dried beans. Fold up the bottom edges to seal and tape across.

PAPER PAWS

Draw the shape of a leopard's paw on two paper bags and paint them yellow. Add big black leopard spots and claws with a black marker. Wear your fierce leopard paws!

Draw a **LEOPON** a cross between a female lioness and a male leopard. Have lots of fun with your imagination.

CHEETAH WORD GAME

How many words can you make out of the word cheetah? You should be able to make at least five words. Look at the letters and try putting them in different sequences to find new words.

ANIMAL MAP ZONE

On a map of the world find the countries where the animals in the True to Life books live in the wild. Label them with pictures of the animals and their habitats. For example, rainforest, jungle, bamboo forest, savannah, etc.

COLLECTIVE QUIZ

Some groups of animals are described using collective nouns. Here are some from the True to Life books:

A journey of giraffes
A herd of elephants
A pod of hippos
A pride of lions
A crash of rhinos
A leap of leopards
A swift of tigers
A clan of hyenas
A litter of cubs

How many collective nouns can you add?

ANIMAL SUPERSTARS

Look for TV advertisements featuring wild animals and explain why you think that animal has been used for the ad.

SAFARI

Plan a safari in Kenya.
What animals might you see?
Where will you stop to camp?
Do you live in a tent?
Describe your guide.
What dangers might you face?
Will you see animals hunting?
How will you travel between camps?
What foods will you eat?
Will you be frightened at night?

ENDANGERED ANIMALS

Name some of the most endangered animals in the world. What can you do to help save them from extinction? Name some of the charities that help endangered animals.

PANDA POEM *See if you can fill in the missing letters.*

I have a furry coat that is black and white;
It keeps me very warm while I sleep at n - - - -.

If you see me with a plant, it is called bamboo;
I like to eat it all day long, and I like to e - - leaves too.

My paws are very big and my jaw is very strong;
With my friends I like to climb and play, all day l - - -.

colour in the pandas and the bamboo forest where they live

SERENA GEDDES

RED PANDA

SCIENTIFIC NAME *Ailurus fulgens*. Sometimes called fire-coloured cat or the lesser panda.

- The red panda is endangered.
- They live in the bamboo forests in China and the Himalayas.
- Fewer than 2,500 remain in the wild. This is due to loss of habitat and being hunted for their beautiful fur.
- They can live up to 14 years in captivity.
- Red pandas are shy gentle animals.
- They weigh up to 6 kilograms and are up to 63 centimetres long. Their tails can be 47 centimetres long.
- Red pandas eat bamboo shoots. They have a false thumb to grasp the stems, the same as giant pandas.
- Red pandas lick their fur and groom themselves, just like a cat.

INTERESTING WEBSITES

GLOBIO www.globio.org

ANIMAL PLANET
http://animal.discovery.com

KIDS' PLANET
www.kidsplanet.org

NATIONAL GEOGRAPHIC
www.kids.nationalgeographic.com/kids

THE SMITHSONIAN NATIONAL ZOO
hhtp://pandas.si.edu

WORLD WILDLIFE FUND
www.worldwildlife.org/pandas

BBC WILDLIFE FINDER-PANDA
www.bbc.co.uk/nature/species/
Giant Panda

ANIMAL INFO
www.animalinfo.org

GIANT PANDA
www.giantpanda.com.au

PANDAS INTERNATIONAL
www.pandasinternational.org

RED PANDA
www.enchantedlearning.com

See exciting videos of wild animals from the True to Life Books http://bit.ly/13gVVoY